Project Earth Science: Astronomy

This book has been edited and produced by the staff of NSTA Special Publications.

Library of Congress Catalog Card Number 92–60246

Stock Number PB–90

ISBN Number 0-87355-108-7

Printed in the United States of America

Project Earth Science: Astronomy

by P. Sean Smith

A Project of Horizon Research, Inc.
Materials for middle school teachers in Earth science
This project was funded by BP America, Inc.

Project Earth Science: Astronomy

Table of Contents

Acknowledgements 7
Overview 9
Introduction 9
Getting Ready for Classroom Instruction 11
Organization 12

Activities

Activity 1 It's Only a Paper Moon 15
Activity 2 Time Traveler 23
Activity 3 Solar System Scale 29
Activity 4 Hello Out There! 37
Activity 5 How Far to the Star? 45
Activity 6 Solar System Soup: The Formation of the Solar System 53
Activity 7 The Goldilocks Effect or "This Planet Is Just Right!" 59
Activity 8 The Greenhouse Effect 67
Activity 9 Creature Feature 75
Activity 10 Reason for the Seasons 83
Activity 11 Ping-Pong Phases 91

Readings

Angular Diameters .... 103 **Reading 1**

What Is a Light Year? .... 106 **Reading 2**

Hubble Space Telescope .... 108 **Reading 3**

Scale Measurements .... 109 **Reading 4**

Scouting Earth/Moon .... 111 **Reading 5**

The Parallax Effect .... 112 **Reading 6**

Understanding the Earth as a System .... 114 **Reading 7**

After the Warming .... 116 **Reading 8**

Grand Theme 2: Atmosphere, Oceans, Cryosphere, and Hydrologic Cycle .... 122 **Reading 9**

The Greenhouse Effect .... 124 **Reading 10**

Grand Theme 4: Interaction of Human Activities with the Natural Environment .... 127 **Reading 11**

Reason for the Seasons .... 129 **Reading 12**

Phases of the Moon .... 132 **Reading 13**

Understanding the Moon Illusion .... 134 **Reading 14**

Appendix

Master Materials List 138

Annotated Bibliography 140

About the Cover Photographs (Reprinted courtesy of NASA)

Front Cover

The photo of Earth rising was taken by the Apollo 10 astronauts, approximately 160,000 kilometers from Earth. Europe, Asia, and parts of Africa are visible through scattered cloud cover.

The photo of the Sun, taken in 1973 from Skylab 4, shows one of the most spectacular solar flares ever recorded, spanning more than 588,000 kilometers across the solar surface.

Back Cover

This color-enhanced photo of the Sun was also taken in 1973 from Skylab. The Sun's corona stretches far beyond the denser, inner corona seen in X-ray and ultraviolet light images, and beyond the limits of what we normally see in the dark sky of a total solar eclipse.

Acknowledgements

Numerous people have contributed to the development of *Project Earth Science: Astronomy*. The volume began as a collection of activities and readings for Project Earth Science, a teacher enhancement project funded by the National Science Foundation. Project Earth Science was designed to provide in-service education for middle school Earth science teachers in North Carolina. Nine two-person leadership teams received extensive training in conducting in-service workshops on selected topics in astronomy, geology, meteorology, and oceanography. They in turn conducted in-service education programs for teachers throughout the state of North Carolina. Principal investigators for this project were Iris R. Weiss, president of Horizon Research, Inc.; Diana Montgomery, research associate at Horizon Research, Inc.; Paul B. Hounshell, professor of education, University of North Carolina-Chapel Hill; and Paul Fullagar, professor of geology, University of North Carolina-Chapel Hill. Brent Ford, president of Novostar Designs, Inc. and a consultant to Horizon Research, also played a major role in this project both in development of activities and organization of workshops.

The activities and readings have undergone many revisions as a result of the comments and suggestions provided primarily by the Project Earth Science leaders and also by workshop participants, project consultants, and project staff. The Project Earth Science leaders made this book possible through their creativity and unceasing attention to the needs of students and classroom teachers. The leaders are: Kevin Barnard, Winston-Salem/Forsyth County Schools; Kathy Bobay, Charlotte-Mecklenburg Schools; Pam Bookout, Guilford County Schools; Betty Dean, Guilford County Schools; Lynanne (Missy) Gabriel, Charlotte-Mecklenburg Schools; Flo Gullickson, Guilford County Schools; Michele Heath, Chapel Hill/Carrboro Schools; Cameron Holbrook, Winston-Salem/Forsyth County Schools; Linda Hollingsworth, Randolph County Schools; Geoff Holt, formerly of Wake County Schools; Kim Kelly, Chapel Hill/Carrboro Schools; Laura Kolb, Wake County Schools; Karen Kozel, Durham County Schools; Kim Taylor, Durham County Schools; Dana White, Wake County Schools; Tammy Williams, Guilford County Schools; and Lowell Zeigler, Wake County Schools. Special thanks to Geoff Holt for his contributions to "How Far to the Star?" and "Ping-Pong Phases," to Kevin Barnard for his contributions to "Solar System Soup," and to Missy Gabriel for her contributions to "Creature Feature."

The manuscript was reviewed for scientific accuracy by Wayne Christiansen, professor of astronomy at the University of North Carolina-Chapel Hill. Dr. Christiansen contributed valuable suggestions on all the activities and the original idea for some of them.

Shirley Brown, a teacher with the Columbus City Schools in Ohio and a participant in the Program for Leadership in Earth Systems Education (PLESE) at Ohio State University, contributed many of the items in the Appendix as well as the poems at the beginning of each activity. She also provided many of the ideas in "Suggestions for Interdisciplinary Reading and Study" in the activities.

During the publication process excellent reviewers provided their comments and suggestions to NSTA: Darrel Hoff, director of Project ESTEEM (Earth Science Teachers Exploring Exemplary Materials) and professor emeritus at the University of Northern Iowa; Mary Kay Hemenway, senior astronomy lecturer at the University of Texas-Austin and education officer for the American Astronomical Society; Katherine Becker, science education specialist at Creighton University; and Mark Brockmeyer, science teacher at City High School in Iowa City.

Project Earth Science: Astronomy was produced by NSTA publications, Shirley Watt Ireton, managing editor; Christine Marie Pearce, assistant editor; Andrew Saindon, assistant editor, Gregg Sekscienski, production editor; Anne Marie Calmes, editorial assistant; and Daniel T. Shannon, editorial assistant. Christine Pearce was the NSTA editor for *Project Earth Science: Astronomy*. Illustrations were created by the author and by Max Karl Winkler. The book was designed by Marty Ittner of AURAS Design.

Special thanks go to British Petroleum America, Inc., for providing the funds to make this book possible.

Overview

Project Earth Science: Astronomy is the first in a four-volume series of Earth science books. The remaining three are *Meteorology*, *Physical Oceanography*, and *Geology*. Each volume contains a collection of hands-on activities developed for the middle/junior high school level and a series of background readings pertaining to the topic area.

Project Earth Science was funded by the National Science Foundation in 1989. Originally conceived as a program in leadership development, the purpose of the project was to prepare middle school teachers to lead workshops on topics in Earth science. These workshops were designed to help teachers convey the content of Earth science through the use of hands-on activities. With the help of content-area experts, concept outlines were developed for each topic area, and then activities were chosen which illustrated the concepts. Some activities were drawn from existing sources, and many others were developed by Project Earth Science leaders and staff.

Over the course of the two-year project, the activities were field tested both in workshops and in classrooms and were extensively revised. After each revision, the activities were reviewed by content experts for accuracy. Finally, all activities were organized in a standard format. During the publication process, the book underwent another extensive review and revision.

The grand theme of the Project Earth Science materials is the uniqueness of Earth among all the planets in the Solar System. Concepts and activities were chosen that elaborate on this theme. The *Astronomy* volume focuses on planetary astronomy. The object of this volume is to give students a sense of viewing Earth from some point beyond the Solar System. By placing Earth in the context of the Solar System and viewing it as "just another planet," it is hoped that students will begin to grasp the unique aspects of the planet. Chief among these aspects is that Earth is the only planet in the Solar System capable of sustaining life.

Introduction

Everyone has wondered if there is life elsewhere in the Universe. Could those distant, tiny points of light seen in the night sky have orbiting planets that might also have intelligent life forms? What is it about planet Earth that makes life possible? Do any other planets in our Solar System have the same or similar conditions? Do they have water or oxygen? Where are the other planets located in relation to Earth and the Sun?

In the last 40 years space exploration has advanced our knowledge of the neighboring planets that share our Sun. Several Mariner spacecraft studied Venus extensively and the unmanned Soviet spacecraft, Venera, actually landed there in 1967. Today the NASA spacecraft Magellan continues to explore Venus with more advanced instruments. The Viking mission landed a spacecraft on Mars and retrieved and analyzed its soil for traces of life. The Voyager missions, launched in 1977 and still continuing, have given us revealing photographs and detailed information on the giant planets of Jupiter, Saturn, Uranus, and Neptune. For the first time we can compare their geological and meteorological conditions with those of Earth. As scientists accumulate this new data, it also reveals how life, as we know it, is unique to Earth.

The space program has given us a snapshot of our own blue and white planet as photographed by journeying spaceships. Imagine what you would see if you could look at Earth from a point above the entire Solar System. What would Earth look like, and how big would it be compared to the other planets? Does it spin and revolve like the others? Do the other planets have a Moon like Earth's? Is the Solar System crowded?

To answer these questions, we must first know where Earth is in relation to all the other planets. The Sun is at the center of the Solar System with fast-orbiting Mercury in the first planet position. Venus is next, followed by Earth and Mars as you move out from the Sun. Moving still farther away are the giant planets Jupiter, Saturn, Uranus, and Neptune. The Solar System's smallest planet, Pluto, is at the time of press closer to the Sun than Neptune even though its orbit extends the farthest of all the planets. This is because Pluto's elliptical orbit slightly overlaps that of Neptune's. Pluto will remain inside Neptune's orbit until March 14, 1999 when it will cross over once again to take its place as the last planet in our Solar System.

Only when Earth is placed in the context of the Solar System and considered as just another planet do its unique features come to light. NASA is in the initial stages of a "Mission to Planet Earth," the goal of which is to study Earth as one global system. This is the perspective NASA has taken in studying every other planet. Now it is turning its attention back to our planet instead of out to the rest of the Universe. Some of Earth's unique aspects have already been discovered using this approach.

As NASA recently wrote: "If Earth were much smaller, it could not retain an atmosphere. If it were much closer to or much further from the Sun, the oceans would boil or freeze. If its orbit and axis of rotation did not fluctuate, the cyclical variations in climate that have spurred evolution would not exist."

Getting Ready for Classroom Instruction

The activities in this volume are organized under three broad concepts. First, students investigate techniques that are used to measure distances and sizes of the magnitude found in the Solar System. Using the information gained from these methods, students then place Earth in the Solar System in relation to the rest of the planets. Second, students perform activities that stress the uniqueness of Earth. This section focuses on comparisons of Earth to other planets, particularly Venus and Mars. In the third section of the book, students are confronted with two areas of planetary astronomy where many people have misconceptions: the reason for the phases of the Moon and the explanation of Earth's seasons.

We suggest organizing Project Earth Science activities around key concepts. This organization works best if used with students or in a workshop for teachers. The presentation and explanation of concepts and the participation in activities should be an integrated process. To facilitate this approach, the conceptual outline for *Astronomy* is presented below with a list of the activities that pertain to each concept.

1. While Earth's position relative to the Sun and other planets is always changing, its *distance* from the Sun is almost constant. To understand many features of Earth which make it unique and habitable, we need to know where it is in relation to the Sun and other planets. Several tools are available to learn where Earth is in the Solar System.

Activities:

It's Only a Paper Moon

Time Traveler

Solar System Scale

Hello Out There!

How Far to the Star?

2. Earth's position in the Solar System is responsible for its unique properties. Among these properties is the fact that Earth is the only planet in the Solar System that sustains life.

Activities:

Solar System Soup: The Formation of the Solar System

The Goldilocks Effect or "This Planet Is Just Right!"

The Greenhouse Effect

Creature Feature

3. Many misconceptions exist about Earth and its characteristics. Looking at Earth as "just another planet" can help correct these misconceptions.

Activities:

Reason for the Seasons

Ping-Pong Phases

Organization

The activities in this volume are designed to be hands-on. In collecting and developing these activities, effort was made to use materials that are either readily available in the classroom or inexpensive to purchase. Novostar Designs, Inc. is offering kits and materials to accompany *Project Earth Science: Astronomy*. For further information, contact Novostar Designs, Inc., 111 West Pine Street, Graham, NC 27253. (919)229–5656.

Each activity has a student section and a teacher guide. The student section has background information which briefly explains the concepts behind the activity in non-technical terms. Following this is the procedure for the activity and a set of questions to guide the students as they draw conclusions.

The teacher versions of the activities, entitled "Teacher's Guide to Activity" contain a more detailed version of the background information given to the students and a summary of the important points that students should understand after completing the activity. You'll find the approximate time to allow for each activity in the section entitled "Time Management." The "Preparation" section describes the set up for the activity and gives sources of materials. "Suggestions for Further Study" gives ideas for challenging students to extend their study of the topic. Use these ideas within the class period allotted for the activity if time and enthusiasm allow. "Suggestions for Interdisciplinary

Reading and Study" includes ideas for relating the concepts in the activity to other science topics and to other disciplines such as language arts and social studies. The final section of the teacher's guide provides answers to the questions in the student section of the activity.

Each activity begins with a poem or quote whose imagery has an astronomical theme. The nighttime skies have been a fascination and an inspiration, to scientific and artistic intellect alike, from time immemorial. You may want to copy these poems to accompany the activities—maybe to remind students of the interwoven nature of arts and sciences or to use elsewhere in the curriculum.

The activities are followed by a section of readings for the teacher. One or more of these readings are referred to in the guide to each activity. The readings provide both background information on the concepts underlying the activities as well as supplementary information to enhance classroom discussions.

Project Earth Science began as a workshop method to provide assistance to teachers on Earth science content and instructional techniques. To use this volume for teacher workshops, we've included a master materials list on page 138.

Lastly is an annotated bibliography of astronomy resources for students and teachers. The bibliography includes activities, curriculum projects, books, audiovisual materials, instructional aids, references, and NASA resource centers. Though not exhaustive, this list should give teachers the means to explore this fascinating subject.

Moon

Moon remembers.

Marooned in shadowed nights,

white powder plastered
on her pockmarked face,
scarred with craters,
filled with waterless seas,

she thinks back
to the Eagle,
to the flight
of men from Earth,
of rocks sent back in space,
and one
faint
footprint
in the Sea of Tranquility.

Myra Cohn Livingston

It's Only a Paper Moon

Background

It is very difficult to measure the diameter of objects in the Solar System. Unlike objects on Earth, it is not possible to hold a meter stick up to the planets and measure them. There are, however, other kinds of measurements called **indirect measurements** that do not require going to the object. These methods allow us to measure objects millions of kilometers away without ever leaving Earth. One type of indirect measurement is known as **angular diameter**.

The easiest way to understand angular diameter is to look at an example. The angular diameter of the Moon is $1/2°$. What does this mean? If it were possible to stretch two strings from the same point on Earth to opposite sides of the Moon, the angle between the strings would be $1/2°$ (see figure 1).

Angular diameters have an interesting property, which you will investigate in this activity, that makes them useful in determining the true size of objects in the Solar System.

Objective

The objective of this activity is to learn how angular diameters can be used to measure the true diameter of the Moon.

Materials

For each group of students:

- ◊ paper plate
- ◊ metric ruler
- ◊ metric tape measure
- ◊ index card
- ◊ scissors

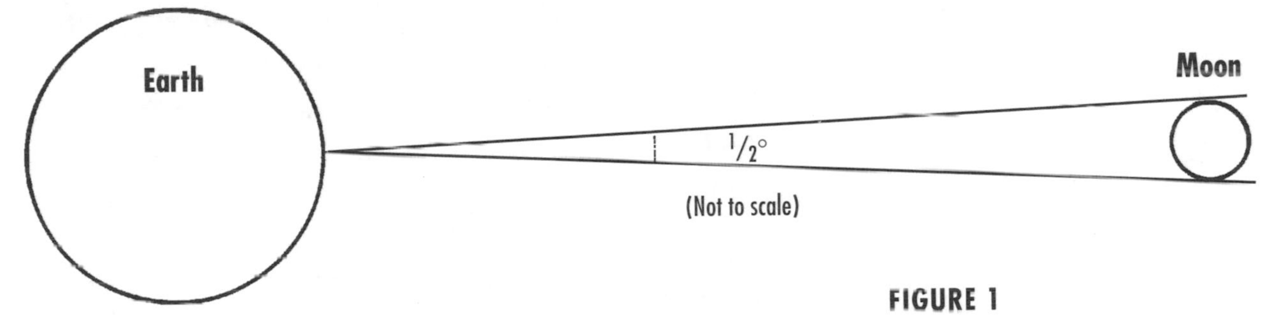

FIGURE 1

Procedure

1. Using scissors, cut a notch and hole in the index card as shown in figure 2.
2. Insert the metric ruler through the index card. On the wall attach the paper plate at eye level or higher.
3. Using the metric tape measure, make sure that your eye is exactly 4 m from the plate. Then, holding the low numbered end of the metric ruler up to your eye, point it at the paper plate and sight through the notch along the ruler at the plate with one eye, as shown in figure 3.

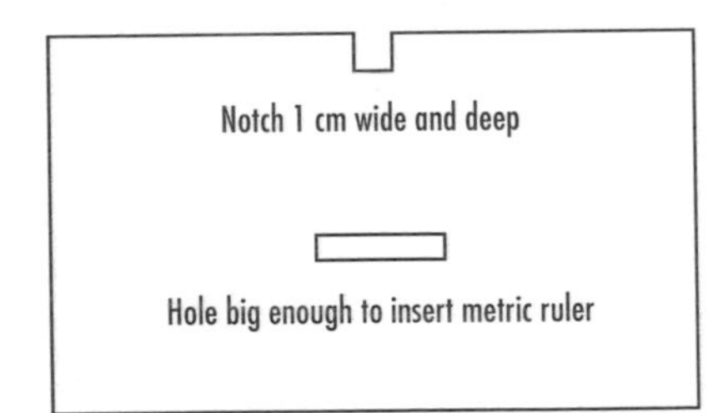

FIGURE 2

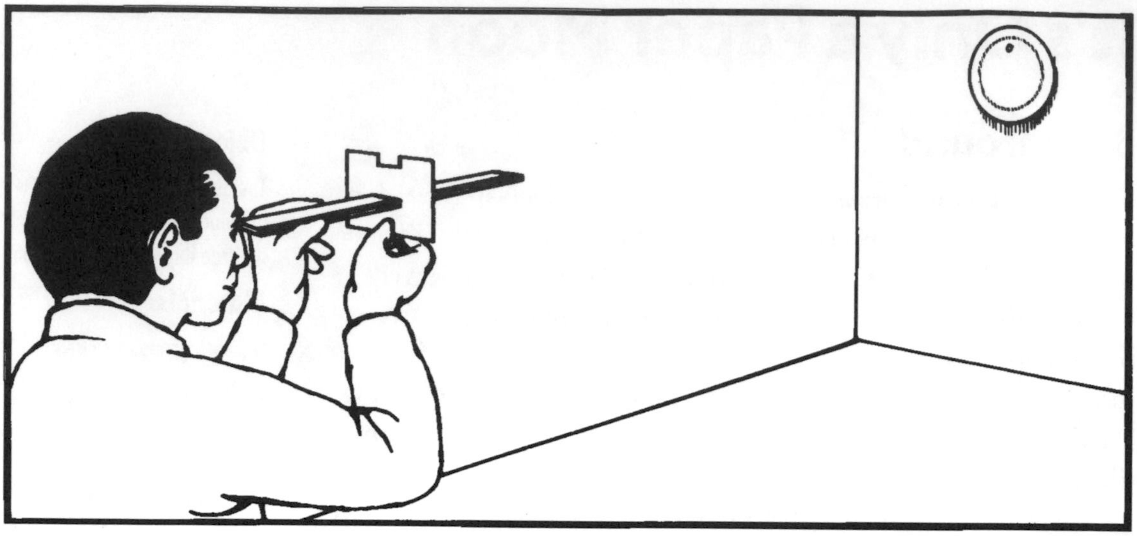

FIGURE 3

4. Slide the index card along the ruler until the paper plate just fits inside the notch. (This is easier to do if you focus on the notch instead of the plate.) Now the notch and the plate have the same angular diameter.
5. Read the distance to the index card and add one centimeter to this measurement to account for the fact that the ruler was not exactly against your eye. Record this in the Data Table on the next page.
6. Repeat step 5 three more times, recording each measurement in the Data Table. Then take the average of the four measurements and record this as well.
7. Using the equation below, calculate the diameter of the plate and record it. This is possible because even though the notch and plate are different sizes, the index card was moved to the point where the notch had the same angular diameter as the plate.

$$\frac{\text{Diameter of the plate}}{\text{Diameter of the notch}} = \frac{\text{distance to the plate}}{\text{avg. distance to card}}$$

$$\text{Diameter of the plate} = \frac{400\text{ cm}}{\text{average distance to card}} \times 1\text{ cm}$$

8. Measure the true diameter of the plate and record it.

Optional: Try this activity at night when the Moon is full and the sky is clear. Everything is done the same way except you will measure the diameter of the Moon instead of a plate. You will probably need a flashlight and someone to help you. You will also need to know the distance to the Moon: 38,440,100,000 cm.

DATA TABLE						
	Distance to card (cm)	**Avg. distance to card (cm)**	**Diameter of notch (cm)**	**Distance to plate (cm)**	**Calculated diameter of plate (cm)**	**Measured diameter of plate (cm)**
Trial 1						
Trial 2						
Trial 3						
Trial 4						

Questions and Conclusions

1. Compare the true diameter of the plate with the diameter you calculated. How do the two compare?
2. If the true and calculated diameters are not the same, what reasons could explain the difference?
3. Would this method of determining diameters be helpful in working with the planets in the Solar System? Why?
4. The full moon appears to be bigger when it is on the horizon than when it is high up in the sky. How could you use the method in this activity to determine whether this effect is real or an illusion?
5. Write a procedure for using the instrument in this activity to measure some tall object on the school grounds. You might choose a flagpole, a bus, or the school building. Be sure that your procedure is specific enough that anyone else reading it could do the measurement themselves.

It's Only a Paper Moon

Materials

For each group of students:

- ◊ paper plate
- ◊ metric ruler
- ◊ metric tape measure
- ◊ index card
- ◊ scissors

Vocabulary

Indirect measurement: A measurement made by other than direct means. Physical proximity to the object or distance is not necessary. Examples of indirect measurement include angular diameter and parallax.

Angular diameter: The measure of an object's diameter in degrees of an arc rather than in linear measurement units. This is a type of indirect measurement. Angular diameters are used to calculate diameters of objects in the Solar System.

Similar triangles: Two triangles that have equal angles but sides of different lengths.

What Is Happening?

Objects in the Solar System are extremely difficult to measure, primarily because they are so far away. There are, however, indirect ways of measuring these objects; for example, measuring angular diameters. An angular diameter is simply a diameter measured in degrees of an arc rather than centimeters or kilometers. The angular diameter of the Moon is $1/2°$, as shown in figure 4.

Angular diameters have several interesting properties. The magnitude of an angular diameter depends on where the object is located. The closer an object is to the observer, the larger its angular diameter. Also, two objects with different *true* diameters can have the same *angular* diameter. If you hold a pencil in front of one of your eyes, you can use it to just block your view of a telephone pole. When you do this, you are adjusting the position of the pencil so that it has the same angular diameter as the telephone pole.

When two objects have the same angular diameter, a principle known as similar triangles can be employed to determine the true diameter of one of the objects as long as the true diameter of the other object is known. In similar triangles, the corresponding angles in each are equal even though the corresponding sides may not be, as shown in figure 5. Although the corresponding sides are not equal, their ratios are. So as long as some measurements are known, others can be determined by setting up the appropriate ratios.

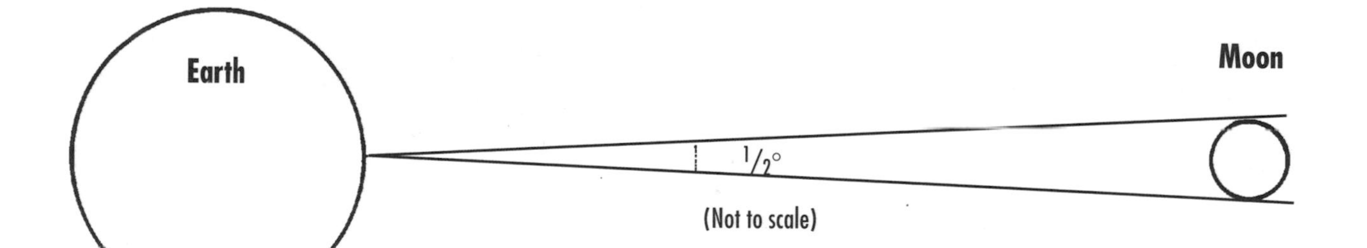

FIGURE 4

Figure 6 shows how the principle of similar triangles and angular diameters can be used to determine the true diameter of the Moon. The distance to the card, the distance to the Moon (the paper plate in this activity), and the diameter of the notch are all known. This means that the true diameter of the Moon can be calculated using the ratio:

$$\frac{\text{diameter of the Moon}}{\text{diameter of the notch}} = \frac{\text{distance to the Moon}}{\text{distance to the card}}$$

In this activity, students will learn how to use this method. For the activity to be more than an exercise in plugging numbers into an equation, it will be necessary to explain to the students the properties of angular diameters that make the equation possible, as described above. These properties are summarized in the next section.

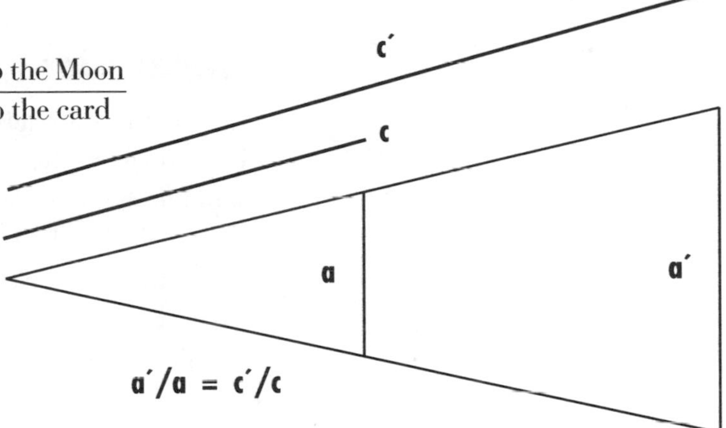

FIGURE 5

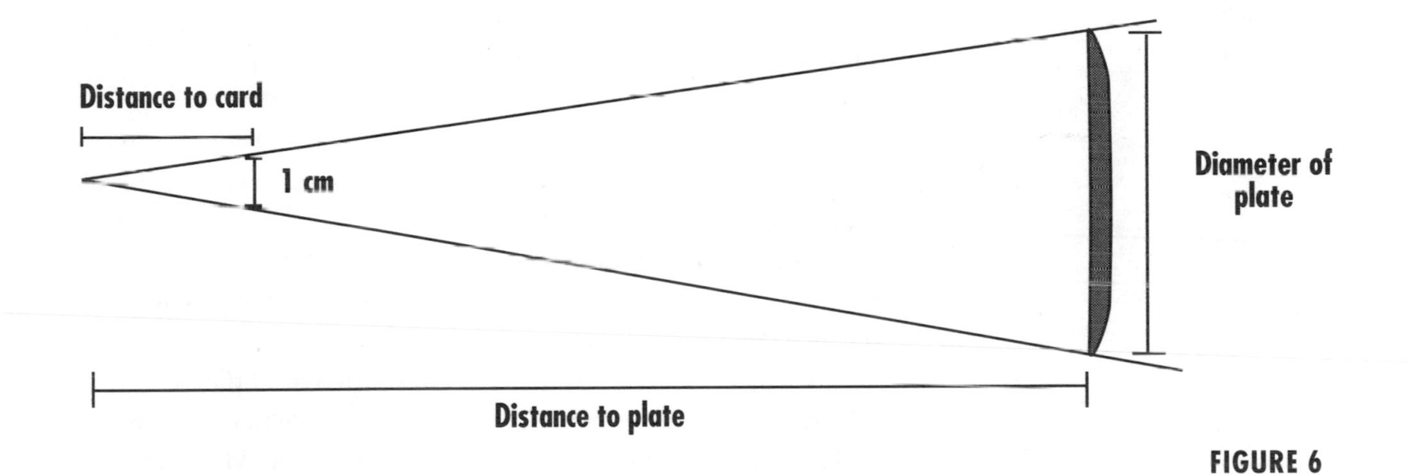

FIGURE 6

Important Points for Students to Understand

◊ While an object's true diameter is constant, its angular diameter depends on the location of the object in relation to the viewer; i.e., the closer an object is, the larger its angular diameter.

◊ Two objects of different size can have the same angular diameter.

◊ When two objects of different size have the same angular diameter, the principle of similar triangles can be used to determine the size of one of the objects.

Time Management

This activity will take one class period or less.

Preparation

Have all the materials centrally located so that students can obtain them easily. Make sure that there is enough space in the room for all the students to work at once, either individually or in groups. It is helpful to mark in advance the spaces where groups of students can work.

For additional background information refer to Reading 1, "Angular Diameters" and Reading 14, "Understanding the Moon Illusion."

Suggestions for Further Study

Encourage students to carry out the optional activity explained in the student section. In addition, they may want to try the present activity with other objects besides paper plates, or they may want to try measuring the diameter of the plate again while standing at a different distance.

Now that Voyager 2 has reached all the planets but Pluto, much more accurate planetary data is available. Students might find it interesting to research some of the findings resulting from the Voyager mission. For example, they could compare Earth science or astronomy textbooks written before and after Voyager reached Neptune.

Have students measure the north/south axis of the Moon during the day. The Moon's first quarter is visible in the morning and its phases from second quarter to full moon are visible in the afternoon. There is a common misconception that the Moon is only visible at night. Students who seldom look will occasionally notice the Moon in the day and think it a fluke without contemplating why.

Suggestions for Interdisciplinary Reading and Study

As a preparation for this activity or as an independent activity, students could write their own definitions of "close" and "far," giving examples of each. A class discussion of the definitions might emphasize for the students how different people can have very different ideas about the same words. This would also be a good opportunity to point out that effective communication in science (and in general) depends both on using well-defined terms and on all parties agreeing on the definitions.

Answers to Questions for Students

1. Depends on students' data. In all likelihood, the two will be close but not the same. Either may be larger than the other, providing a good opportunity to explore sources of error in measurement.
2. Among the many sources of experimental error are: inaccurate measurement (of plate diameter, of index card distance or notch diameter, of distance to plate), not remaining stationary from measurement to measurement, and changing position of the metric ruler from measurement to measurement.
3. This method, if adapted for use with a telescope, works well for all the planets except Pluto. (Pluto is too far away and too small, even if you are using a telescope.)
4. Calculate the Moon's diameter at moonrise and at "moon noon." Compare the two. They should be the same.

   The reason the Moon appears bigger on the horizon is that we perceive the horizon to be farther away than the sky above us. So for the Moon to have the same angular diameter at two different distances (on the horizon and above us), our mind interprets the object farther away (on the horizon) as being larger than the one nearer (overhead) to us. (Refer to the Reading 14, "Understanding the Moon Illusion," for further explanation of this phenomenon.)
5. ***Note*:** Answers will vary. The important thing is that the procedure is feasible. If time and materials are available, encourage the students to try their own procedures. There could even be a class vote to determine the best procedure based on feasibility, clarity, and accuracy of results.

A Distant Sun

A distant sun is like a dream,
so very far away,
through clouds of doubt and darkness
is very hard to see.

But, ah, we feel its beauty,
though filtered it may be,
rays of hope and promises
light up our shadowed ways.

As we watch it cross the sky,
through silhouettes of life,
we feel our distant dreamings
draw closer to our sight.

Lydia Ferguson

Time Traveler

Background

Distances in astronomy are often very difficult to comprehend because they are so large. For example, the distance from the star Sirius to Earth is 84,320 trillion km. This distance is too large for most people to imagine or understand. There are ways, however, to make such large numbers more manageable. For example, it is much easier to understand and work with 15 years than with 5,475 days, even though they both represent the same amount of time. A "year" is just a much larger unit than a "day." The same type of thing can be done with distances using a measurement known as a **light year**.

A light year (abbreviated "ly") is a measurement of distance even though it involves a time unit, the year. A light year is defined as the distance that light will travel in one year. The speed of light is 300,000 km/s. To calculate how far light travels in one year, first calculate how many seconds there are in a year:

$$60 \text{ s/min} \times 60 \text{ min/hr} \times 24 \text{ hr/d} \times 365 \text{ d/yr} = 31{,}536{,}000 \text{ s/yr}$$

So in one year, light will travel

$$31{,}536{,}000 \text{ s} \times 300{,}000 \text{ km/s}$$
$$= 9{,}461 \text{ trillion } (9{,}461{,}00{,}000{,}000{,}000) \text{ km}$$

This is the same as traveling around the world 118 million times.

The light year can make distances easier to understand in the same way that a year makes a large number of days more understandable. Returning to our example above, the distance to the star Sirius from Earth is 84,320 trillion km. But this distance is only 9 ly, a much more manageable number.

The light year is also important because it tells us about the time lag involved in communicating over the large distances involved in astronomy. If we sent a television or radio signal (both of which travel at the speed of light) to Sirius, it would be nine years before it arrived there. In the same way, if Sirius were to stop shining right now, we would not find out about it for nine more years, when the last light the star produced finally reached Earth. How old would you be then?

Objective

The objective of this activity is to develop an understanding of the concept of a unit known as the "light year."

Materials

For each group of students:

- ◊ watch with second hand or a stop watch
- ◊ metric tape measure (30 meters or more works best)
- ◊ calculator (helpful but not essential)

Procedure

1. Find a long distance that you can use either inside or outside the school. This could be a long hallway, the cafeteria, a

parking lot, or a football field. You won't need a distance longer than a football field.

2. Starting at one end of the space you have chosen, walk heel-to-toe for exactly one minute. Mark where you stop.
3. Using the tape measure, measure how far you walked to the nearest meter. Record this distance in the Data Table.
4. Repeat steps 2 and 3 three more times.
5. Calculate the average of the four measurements and record it in the Data Table.

DATA TABLE

	Distance walked	Average distance walked
Trial 1		
Trial 2		
Trial 3		
Trial 4		

6. The average you calculated is the distance you can walk heel-to-toe in one minute. We will call this distance a "student minute."

Questions and Conclusions

1. Are all the student minutes the same? How are they similar?
2. How are student minutes similar to a light year? How are they dissimilar?
3. How many meters are in 3 student minutes?
4. How many of your student minutes are there in 5,000 m?
5. Listening to the radio one morning at 6:30, you hear that school has been cancelled because of damage done in some parts of town by a windstorm. You start to climb back in bed and sleep the day away, but then you remember that your best friend lives in a part of town which was heavily damaged and has no telephone or electricity. Your friend leaves for school every morning at 7:00 and lives 900 m away from you. The only way you can get the news to your friend is to go to your friend's house and deliver the message. If you are only allowed to walk heel-to-toe, can you make it to the house in time based on your own student minute? Explain.

Time Traveler

What Is Happening?

Distances in astronomy are extremely hard for adults to comprehend, let alone middle or high school students. The units typically used for large measurements on Earth (the kilometer or mile) are too small to be of much help when measuring the distance to stars or other galaxies. For this reason, a unit known as the light year was developed. A light year is defined as the distance that light can travel in one year. Traveling at 300,000 km/s, light travels almost 9,461 trillion km in one year.

The light year is a difficult unit for students to understand. The problem lies in the terminology. A time unit, the "year," is being used to measure distance. This is not at all unfamiliar to students, however. It is very common today to talk about distance in terms of time. Students often do not talk about how many miles it is to a friend's house but how many minutes or hours it takes to get there. A comparison of the time-distance to a friend's house when walking, bicycling, or riding in a car can be a useful illustration of the concept that the time-distance unit depends on the speed of the carrier. Since light is the fastest (and most constant) possible carrier, it is the most useful for establishing the scale of huge astronomical distances.

The light year has another important property that is discussed in more detail in Activity 4, "Hello Out There!" The distance in light years that an object is from Earth is the amount of time in years that light from this object would take to reach Earth. For example, if a star is 12 ly from Earth, light from the star takes 12 years to reach us. Likewise, any light from Earth takes 12 years to reach the star.

This activity is designed to help students understand the light year by creating for themselves a unit which is similar, one that uses time to measure distance. It is important that students understand the nature of the unit. Creating their own unit should aid in this understanding. An important distinction, however, needs to be made between the "student minute" and the light year. The student minute may vary depending on how fast the student walks. The light year does not vary. It is a constant because the speed of light is a constant.

Materials

For each group of students:

- ◊ watch with second hand or a stop watch
- ◊ metric tape measure (30 meters or more works best)
- ◊ calculator (helpful but not essential)

Vocabulary

Light year: A unit of measurement equal to the distance light travels in one year (9,461 trillion km or 9,461,000,000,000,000 km).

Important Points for Students to Understand

- ◊ Distances in astronomy are too large to work with easily when expressed in units such as kilometers. Using a larger unit makes these numbers easier to manage.
- ◊ The "light year" is a measurement of distance even though it involves a time unit.
- ◊ It is very common to express distances in terms of time.
- ◊ Light has a finite speed. It takes time for light to travel over any distance.

Time Management

Depending on the availability of watches and tape measures, this activity should take one class period or less.

Preparation

You should determine in advance what area the students will use for this activity. If an outdoor site is to be used, be sure the conditions are such that the students can remain outside comfortably for an extended period of time.

Rather than having the students measure each time they walk, it may be possible to use a lined football field. Alternatively, the teacher may mark off the distance. This can be done simply with a tape measure and marking paint.

For additional background, see Reading 2, "What is a Light Year?" and Reading 3, "Hubble Space Telescope."

Suggestions for Further Study

It is helpful for students to gain experience with conversions. One way to accomplish this is to have them convert some distances they are familiar with to their own student minute. They may also find it interesting to convert the distance to some stars, which are recorded in light years, into kilometers.

Students can also determine their student minute doing other things besides walking heel-to-toe. They may walk normally or run.

Suggestions for Interdisciplinary Reading and Study

The light year is simply a unit that makes vast distances easier to work with, as explained in the Background section. There are examples of units like this in students' everyday lives. For example, a "dozen" is just another way of representing twelve of something. It is easier to talk about ten dozen eggs than 120 of them. An hour is just an easier way of representing 60 minutes. Ask students to think of other units like this in their own experience. In a writing activity, they could invent their own unit, name it, describe it, and explain how it would be used. For example, what is a "mom minute," as in "Yes, Mom. I'll be there in a minute!"

The concept of distance in the Solar System is sometimes expressed in literature. One example is the poem at the begining of this activity, "A Distant Sun," by Lydia Ferguson.

Answers to Questions for Students

1. No, all the student minutes should not be the same.
2. The two are similar in that they both involve a time unit in a distance measurement. Both involve measuring how far something can travel in a given amount of time. The major dissimilarity is in the size of the respective units. But also, the light year is a constant because the speed of light is constant. The student minutes will vary because of the variation in how fast different individuals walk as well as variation in how fast one person walks at different times.
3. Again, this answer will depend on the student minute. For example, if a student minute is calculated to be 30 m, then the answer is:

   (3 student minutes) × (30 m/student minute) = 90 m.
4. This answer will vary depending on each different student minute. A student minute of 30 m is reasonable to assume. In this case, the answer would be:

   5,000 m ÷ 30 m/student minute = 166.7 student minutes.
5. Assuming 30 m for the student minute, it would take the student 900 m/30 m = 30 student minutes to reach the friend's house. Therefore, the student should get there just as the friend is leaving.

The Planets

Across
your dark ocean
the chilled planets journey.
Wanderers of night, they travel
the paths

of their
curving orbits:

Tiny Mercury, Mars,
Venus, bright-ringed Saturn and Earth,
Neptune,

Pluto,
Uranus and
Jupiter, swathed in his
many-colored cloak, his red eye
flashing.

Myra Cohn Livingston

Solar System Scale

Background

Sizes and distances in the Solar System are difficult to visualize. The distance from the Sun to Earth is 150 million km. The diameter of Jupiter is 140 thousand km. Both of these measurements are so much larger than anything you ever see that they are difficult to imagine. But there is another way of thinking about the Solar System that is much simpler. It involves reducing all the sizes by the same amount: for example, dividing all the sizes and distances by two. These new values can be used to make what is known as a **scale model**.

Examples of scale models are all around. Model railroads are scale models of trains. A globe is a scale model of Earth. Figure 1 on page 31 shows a scale model of the relative sizes of the planets, but their relative distances are not drawn to scale. The advantage of scale models is that they allow us to determine the distance and size of the true object. All that is needed is the **scaling factor** that was used in making the model. For example, if the wheels of a model car are 10 cm in diameter, and the wheels of a real car are 70 cm, then the scaling factor is $70 \div 10$ or 7. Now, any size in the real car can be determined by looking at the model car. If the door of the model is 20 cm long, then the door of the real car is 20×7 or 140 cm long.

Johannes Kepler built a scale model of the Solar System almost 300 years ago using the best estimates for size and distance available at his time. As his base scale, he used what would later become known as the Astronomical Unit, the distance between the center of mass of the Sun and the center of mass of the Earth-Moon system. Once the true length of an AU was found (150 million km), the scaling factor could be determined and the rest of the distances calculated.

Objective

The objective of this activity is to build a scale model of the planetary distances in the Solar System.

Materials

For each group of students:

- ◊ about 100 m of string
- ◊ masking tape
- ◊ 10 flags
- ◊ measuring tape (30 meters or more preferably)
- ◊ marker

Procedure

1. Before starting this activity, picture in your mind what you think a scale model of the Solar System will look like and write a brief description of it. See if the model you build meets your expectations.
2. Measure the longest distance you can use, no more than 100 m. Measure this distance to the nearest meter and record it in Data Table 1. This distance will represent the

distance between the Sun and the planet Pluto (that is 39.4 AU or 5.9 billion km).

3. To calculate the distance from the model sun to each model planet, you need to calculate a scaling factor. Determine the scaling factor by dividing the distance from step 2 above by the distance from the Sun to Pluto. Find this distance in Data Table 2. Record the scaling factor in Data Table 1. For example, if the longest distance usable is 78 m, then the scaling factor is 78 m ÷ 39.3 AU = 1.98 m/AU.
4. Multiply the scaling factor from step 3 by the actual distance from the Sun to each of the planets in AU. Use the distances in Data Table 2. Record the answer in the column labeled "scale distance from Sun."

DATA TABLE 1

Largest usable distance (meters)	Distance to Pluto (AU)	Scaling factor (m/AU)
	39.4	

DATA TABLE 2

Planet	Distance from Sun (AU)	Distance to planet (kilometers)	Scale distance from Sun (meters)	Actual diameter (kilometers)
Sun (a star)	n.a.	n.a.	n.a.	1,391,980
Mercury	0.39	58,000,000		4,880
Venus	0.72	108,000,000		12,100
Earth	1.00	150,000,000		12,800
Mars	1.52	228,000,000		6,800
Jupiter	5.20	778,000,000		142,000
Saturn	9.54	1,430,000,000		120,000
Uranus	19.2	2,870,000,000		51,800
Neptune	30.1	4,500,000,000		49,500
Pluto	39.4	5,900,000,000		2,300

5. Measure out a length of string equal to the scale distance to Mercury from the Sun. Do not cut the string, but wrap a piece of masking tape around it at the proper distance and write “Mercury” on it. From that point, continue measuring the same string out to Venus, and mark that spot on the string. Continue doing this for all the planets out to Pluto.
6. Stretch the string out, and then attach a flag to the string at each point where the location of a planet is marked.

Questions and Conclusions

1. Describe what your model looks like. Is this different from what you pictured in your mind in step 1? If so, how?
2. The nearest star to Earth is Alpha Centauri, 274,332 AU away. Where would this star be placed in your scale model of Solar System distances?
3. What are some of the advantages and disadvantages that you see in using a scale model? Be specific and use examples from this activity.
4. If you were to make a scale model of the Milky Way Galaxy, what scaling factor might you use?

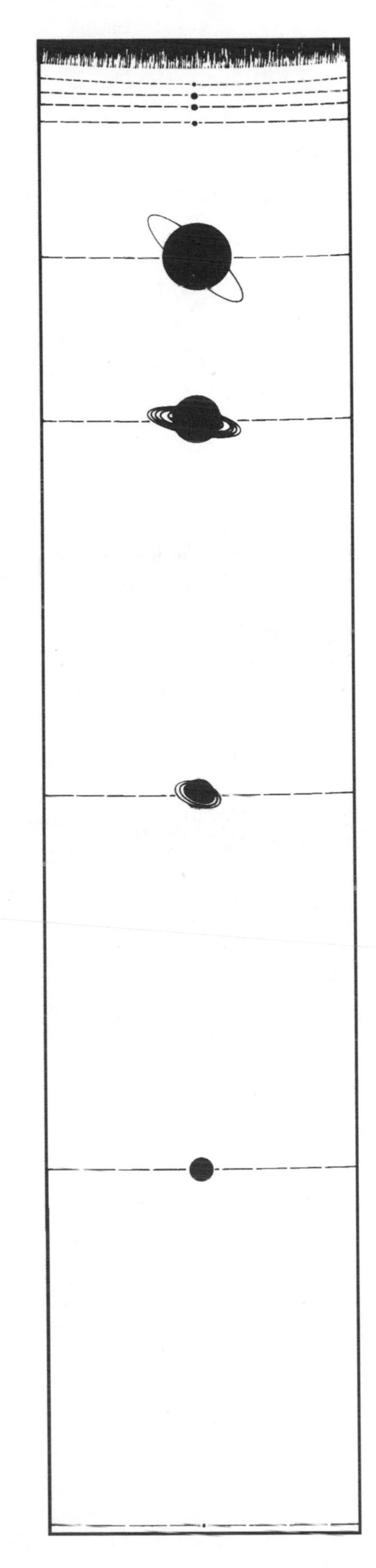

FIGURE 1
(Distances not to scale)

Solar System Scale

Materials

For each group of students:

- ◊ about 100 m of string
- ◊ masking tape
- ◊ 10 flags
- ◊ measuring tape (preferably 30 meters or more)
- ◊ marker

Vocabulary

Scale model: A model that is proportional in all respects to the object being modeled.

Scaling factor: The factor or proportion which, when multiplied by measurements of a scale model, gives the measurements of the object.

Astronomical unit: The basic unit of length used to measure distances in the Solar System. It is the distance from the center of mass of the Sun to the center of mass of the Earth-Moon system (149,600,000 km).

What Is Happening?

Sizes and distances in the Solar System are extremely difficult to visualize. The distance from the Sun to Earth is 150 million km. The distance is so great that there is nothing in everyday experience with which to compare it. Working with distances of this magnitude is extremely difficult for students and adults alike.

One way of dealing with these distances is to use scale measurements, a form of indirect measurement. Many models of buildings and cars use scale measurements. In order to determine true measurements from scale models, one must only know the **scaling factor**. For instance, if a model car is one-seventh the size of the real car on which it is based, then every measurement on the real car is simply the scaling factor, seven, times the corresponding measurement on the model. If the wheels on the model are 10 cm in diameter, the real wheels are 10 cm $\times$ 7, or 70 cm.

In this activity, students will build a scale model of the Solar System using true measurements and a scaling factor. The model involves only the distances to the planets, not their sizes. In order to use the same scaling factor for both distance and size, a huge open area would be required. Without such a large area, the sizes of the planets would be much too small to see easily. This fact emphasizes one of the important features of the Solar System—it is mostly empty space. Students need to be made aware of another shortcoming of this model—it represents the planets as all being aligned on one side of the Sun. In reality, such an arrangement of the planets happens infrequently.

Important Points for Students to Understand

- ◊ The Solar System is largely empty space.
- ◊ A scale model is one way of working with distances that are too large to visualize.
- ◊ As long as the scaling factor is known, true measurements can be determined from scale measurements and vice-versa.
- ◊ The planets are rarely lined up on one side of the Sun.

Time Management

This activity will probably take more than one class period but less than two. One option is to have the students do all their calculations and string measurement on the first day and then to have them actually lay out the full model on the second day.

Activity 4, "Hello, Out There!" uses the same Solar System scale you have just created. It takes less than a period, so you may want to schedule this at the end of the second day.

Preparation

If an outdoor space is being used, it is important to do the outdoor part of the activity on a day when the students will be comfortable outside for a long period of time. This activity can be done indoors either in a long hallway, the gymnasium, the cafeteria, the auditorium, or any other large open space.

For additional background information, refer to Reading 4, "Scale Measurements" and Reading 5, "Scouting Earth /Moon."

Suggestions for Further Study

Encourage students to calculate what the scale size of the planets would be using the same scaling factor as in the first part of the activity. Students may want to pursue the idea of finding a space large enough to lay out a scale model of the Solar System that does have actual models of the planets. An excellent exercise is to have them determine how large that space would be. At the very least, it would be several city blocks long. If such a space is easily accessible, building the model would be very instructive.

Groups of students can be assigned to research individual planets with regard to their important and unique features. Then, once the solar system is laid out, they can conduct a walking tour of the solar system, stopping at each planet to learn about it.

Suggestions for Interdisciplinary Reading and Study

Scale models are used in all aspects of Earth science—meteorology, geology, oceanography, and astronomy. The maps used in these areas can be scale models. In each case, the purpose of the model is to depict something which is very large in a much smaller size or something which is very small in a much larger size (like an atom, molecule, or mineral).

Computer models have been particularly useful in meteorology in studying and predicting weather patterns. Many researchers are now trying to create models to study the greenhouse effect and global warming. They are attempting to use these models to predict the future effects of increased carbon dioxide in the atmosphere. These applications point out some of the advantages and disadvantages of models. Encourage students to investigate the use of models and how these models are different from scale models in other areas of Earth science.

The book *This Island Earth* (see Bibliography) provides descriptions of Earth as one of the nine planets in the Solar System. The accounts in the book emphasize the beauty and uniqueness of Earth among all the other planets (see Reading 5 for one such account).

The Solar System has been a source of inspiration for many authors. This is particularly true in science fiction and in poetry. Three examples of poems based on the Solar System are "The Planets" (beginning of Activity 3) by Myra Cohn Livingston and "A Distant Sun" (beginning of Activity 2) by Lydia Ferguson, and "Jewels" (beginning of Activity 5) by Myra Cohn Livingston.

Answers to Questions for Students

1. The inner planets are much closer together than the outer planets. The distance between planets increases as distance from the Sun increases. The model is mostly empty space.
2. The answer will depend on the scale used by the students. If the scale is 1.98 m /AU, then the distance the star would be placed away from the Sun would be

$$1.98 \text{ m/AU} \times 274{,}332 \text{ AU} = 543{,}178 \text{ m}.$$

3. Answers will vary with students. One disadvantage of the model in this activity is not being able to construct a size model on the same scale as the distance model. One advantage is the ability to represent the distances to the planets in a space that allows students to appreciate the emptiness of the Solar System.
4. Answers will vary with students, but the learning will come in thinking about the relative distances within the Galaxy compared to distances within the Solar System. The diameter of the Milky Way is estimated at 300,000 parsecs. One parsec is 3,260 ly. One light year is 63,000 AU. A visible model of these dimensions is nearly impossible to scale.

Messages

Space sends messages,
mysterious sounds:

Pulsing beats
from distant
neutron stars;

Radio blackouts from a solar flare;

Hissing meteorites
thundering
the
ground;

Signals from puzzling quasars;

Small,
strange
whistlers
from
Jupiter.

Myra Cohn Livingston

Hello Out There!

Background

Everything we see, we see because of light. We see a tree because light reflects off the tree and travels to our eyes. Light travels so fast (300 thousand km/sec!) we have trouble believing that it takes the light any amount of time to travel from the tree to our eyes. For this reason, when we see something happen, we assume that it is happening at that instant. If we go to a baseball game and see a ball being hit, we assume that the ball was hit the moment we saw it happen. But we do not assume the same thing for sound. For example, when we see a bolt of lightning off in the distance, we see the flash of light before we hear the thunder. The light from the lightning reaches us before the sound does because sound travels much more slowly than light (340 m/sec at 15° C).

All forms of light (X-ray, radio waves, microwaves, visible light) travel at the same speed—300 thousand km/sec. Why does light seem to reach us instantly? For two reasons: the distances we are familiar with are very short, and light travels a short distance very fast. When distances become great, the lag time between something happening and us seeing it happen becomes more obvious, just like with the sound of thunder after we see the lightning. For example, the light we are seeing from the Sun actually left the Sun a little more than eight minutes ago.

Stars other than the Sun are much farther away, therefore lag time is much greater. Distances to stars are often measured in light years. A light year is the *distance* that light can travel in one year. If a star is six light years away from Earth, that means that the light leaving the star takes six years to reach us. That also means that the things we are seeing happen on that star actually happened six years ago. In the same way, if there were people on a planet close to the star looking at us, they would see what happened here in 1986, not 1992. They will never be able to see anything more recent than six years in our past.

Objective

The objective of this activity is to understand the consequences of light having a finite speed.

Procedure

This activity uses the model of the Solar System constructed in the activity "Solar System Scale." Your teacher will give you the directions for the present activity.

Questions and Conclusions

1. Did you find out what was happening on Earth the instant it happened? Why or why not?
2. What were the consequences of the message from Earth taking time to reach you?
3. How does what you learned in this activity apply to sending information to and receiving information from spacecraft by radio waves? Remember that radio waves are a form of light.

Hello Out There!

What Is Happening?

In common experience, it is assumed that light travels over any distance instantaneously. This is because the distances we normally encounter are relatively short, and the speed of light is very great. However, light does take time to travel, and when the distances become great enough, this becomes apparent. An analogy can be made with sound, which travels more slowly than light. If the distance is great enough, there is a lag between seeing an event and hearing the sound that results from it. For example, when we see a bolt of lightning off in the distance, we see the flash of light before we hear the thunder. The light from the lightning reaches us before the sound does because sound travels much more slowly than light.

All forms of light travel at the same speed, 300,000 km/s. Radio waves, one form of light, are used to communicate with spacecraft. The signals now being received from Voyager take hours to reach Earth. The star Sirius (the brightest star as seen from Earth except for the Sun) is 9 ly away. This means that the light now reaching Earth from Sirius left the star nine years ago. If there were people on Sirius watching Earth, they would now be seeing what happened here nine years ago. All this is because light takes time to travel.

This activity is designed to help students understand that light does have a finite speed and that this has consequences for us. In order to grasp the meaning of the activity, it is important for students to understand that light acts as a messenger in the same way that a person can. Both transfer information from one point to another. How quickly the information is transferred depends on the speed of the messenger, whether the messenger is a person or light.

Important Points for Students to Understand

- ◊ Light takes time to travel.
- ◊ Distances must be very great in order for the consequences of the speed of light to become apparent.
- ◊ A fundamental consequence of the finite speed of light is that we always see a star's past, not its present.

Materials

For each student:

- ◊ a piece of candy or other treat (make sure no one is allergic to it)

Vocabulary

Speed of light: According to Albert Einstein's theory of relativity, the speed of light is constant: 299,800,000 m/s (usually rounded to 300,000,000 m/s for purposes of calculation). In this book, 300,000 km/s is used for comprehension and comparison.

Speed of sound: This varies depending upon the conditions through which sound moves, such as the density and the temperature of the medium. For the purpose of this activity, the speed of sound at 15° C in air is 340 m/s.

Radio wave: Like visible light, this is a form of electromagnetic radiation ranging in wavelength from 1 cm to 100 km. Radio waves are low in energy and are used for, among other things, communication with spacecraft. Like all electromagnetic radiation, radio waves travel at the speed of light.

Time Management

This activity will take much less than one class period. It may be possible to do it on the same day as day two of the activity "Solar System Scale."

Preparation

This activity should be done in conjunction with Activity 3, "Solar System Scale," or a similar activity on the Solar System. It also assumes that the students are familiar with a light year, as described in Activity 2, "Time Traveler." Assuming some kind of scale model of the Solar System has been built (for example, the one in Activity 3, "Solar System Scale,") there is very little preparation required other than buying the candy for the students. Time may be saved by dividing the students into their planet groups and designating the messenger prior to the activity.

For additional information, refer to Reading 2, "What is a Light Year?" and Reading 3, "Hubble Space Telescope."

Instructions

1. Select a volunteer or designate a student to be the "electromagnetic messenger."
2. Divide the class into three or four groups and position one group at each of the following planets in your scale model of the Solar System: Mars, Saturn, Uranus, and/or Pluto. Have the students face *away* from Earth.
3. Tell the class that you (the teacher) will be on Earth sending out radio messages into the solar system via the electromagnetic messenger.
4. The electromagnetic messenger will walk heel-to-toe. The students must also walk heel-to-toe when they travel.
5. Since sound cannot travel in outer space, there must be no talking.
6. Send the messenger out into the solar system with a card for each planet that is inhabited. Each card should read, "Mr./ Ms. _________ is on Earth handing out a limited amount of candy to hungry astronauts. Go to Earth if you want some candy. Remember, you must walk heel-to-toe, and NO TALKING!" Students must read the message silently.
7. As students arrive, hand each one a piece of candy inconspicuously. When the last group arrives, tease them that you

have run out of candy, but don't forget to give it to them later.

Suggestions for Further Study

Encourage students to come up with a more elaborate skit to demonstrate consequences of light having a finite speed. This would be an excellent opportunity for students to work in cooperative learning groups. Their group work could be an assessment technique to evaluate whether or not the students grasp the concept.

Have students investigate how astronomers deal with the fact that what they see from stars happened years ago. How do they solve the problem of never being able to see what is happening in a star's present?

Suggestions for Interdisciplinary Reading and Study

Light is electromagnetic radiation. It includes infrared, visible, ultraviolet, and X-rays, and all travel at exactly the same speed. The various forms of light have both harmful and helpful effects. X-rays revolutionized diagnostic medicine. Too much X-ray radiation, however, has been shown to be harmful. Students can investigate the uses of all the different forms of light and their effects on humans.

Various forms of light provide valuable information for astronomers about the Universe. The idea behind the Hubble Space Telescope was to place a device to gather light (of both visible and ultraviolet wavelengths) above the atmosphere where it could get an unobstructed picture of these types of light from objects in space. When light travels through the atmosphere, it can be refracted, reflected, or absorbed, making it difficult to get an accurate picture of the light from planets and stars with a ground-based telescope. In orbit around Earth, Hubble receives light that is unaffected by Earth's atmosphere. See Reading 3, "Hubble Space Telescope," for a more detailed description.

The idea of seeing a star's past brings up the topic of time travel. In a sense, we are going back in time when we observe any star because we see the star's past as though it were the present. Time travel is a popular theme in science fiction. Madeleine L'Engle's *A Wrinkle in Time* is an account of travelling through time between the planets. The poem "Messages" by Myra Cohn Livingston at the beginning of this activity describes some of the different messages we receive from space. Encourage students to investigate the phrases "pulsing beats from distant neutron stars" and "radio blackouts from a solar flare" from the poem.

A writing activity that integrates astronomy and geology could be inspired by the following scenario. "A huge mirror has been discovered fifty million light years from Earth in space. With your telescope, you can focus on the mirror and see a reflection of Earth. Your telescope is so powerful that you can make out as much detail as a small river on Earth. Write an essay that describes what you would see happening on Earth and when in Earth's past these things would have happened." (This would be 100 million years ago, as light reflected from the Earth had to travel 50 million years to reach the mirror, then another 50 million to reach the telescope.)

While not apparent in this activity, light can be modeled as a wave (though it also behaves in some ways like a particle). Waves have relevance in other areas of Earth science besides astronomy. Some geologists study waves that travel through our planet, particularly those that are the result of earthquakes. It is these waves that cause damage from earthquakes. Sound is another example of a wave. Oceanographers often study ocean waves. These waves have properties in common with light waves. Encourage students to compare and contrast these different waves.

Answers to Questions for Students

1. No. Because it took time for the messenger to reach each planet. He or she did not get there instantaneously.
2. All students got the message "late." For the farthest students, this meant not getting to Earth before the candy ran out.
3. Any information we send or receive takes time to arrive even at the speed of light. For example, if NASA wants the Voyager spacecraft to make a turn at a certain time, the message must actually be sent hours before that time so that the message will have time to travel.

Jewels

Space blazes with jewels,
a shimmering ice
of billions of diamonds
dazzles
the Milky Way:

Jupiter, a giant agate,

Uranus, a ball of jade,

Pluto, a luminescent pearl,

Saturn, a halo of rings.

A slice
of moon, a crescent brooch.

Bright rubies splay
Antares

in this
midnight
masquerade.

Myra Cohn Livingston

How Far to the Star?

Background

The dimensions of objects can be measured in two ways—directly and indirectly. A tape measure is an example of a device used to measure things directly. **Direct measurements** are made by stretching a tape or placing a ruler next to an object to find out how long it is. Direct measurements are usually made on objects that can be handled. When objects are too big or too far away to be handled, indirect measurements must be made.

There are several forms of indirect measurement. Scale measurements and angular measurements are discussed in Activity 3, "Solar System Scale" and Activity 1, "It's Only a Paper Moon," respectively. Another form uses the parallax effect. This effect is easily demonstrated. Hold up your thumb at arms length, and with one eye closed, line up your thumb with some object in the background. A tree or a telephone pole will work well. Without moving your thumb, switch eyes. Your thumb appears to have moved relative to the objects in the background of your view. This is the **parallax effect**—the apparent movement of an object when viewed against a stationary background from two different points. The two different points in this example were just your eyes. The fact that we have two eyes allows our brain to take advantage of parallax in judging distances.

The parallax effect is often used by surveyors to measure relatively small distances, for example, when building roads. Direct measurements could also be used for these distances but not for distances to objects that are too far away for humans to reach, like the stars. The parallax effect is one of the few ways we have to measure these cosmic distances from here on Earth.

In this activity, you will investigate the factors that affect parallax.

Objective

The objective of this activity is to investigate the factors affecting parallax.

Materials

For each group of students:

- ◊ construction paper (or manila folder)
- ◊ pencil
- ◊ metric ruler
- ◊ chalkboard or large sheets of paper (69 cm X 81 cm)
- ◊ chalk
- ◊ scissors (if manila folders are used)
- ◊ tape (if manila folders are used)
- ◊ single hole punch

1 2 3 4 5 6 7 8 9 10 11 12 13 14 15 16 17 18 19 20

3 cm

FIGURE 1

Procedure

Experiment 1—In this part of the activity, you will determine the effect on parallax of the distance of the object from the observer.

1. Draw a series of equally spaced vertical lines three centimeters apart across the chalkboard and number them in order as shown in figure 1. Twenty of these lines should be enough.

2. Have one student stand 2.1 m from the chalkboard and hold up a pencil at arm's length (their arm should be parallel to the board). Now, you move back away from the person holding the pencil and away from the chalkboard the distances specified in Data Table 1. Be sure that the student holding the pencil stays in the same place.
3. Close your left eye and look at the pencil with your right eye. In Data Table 1, make a mark below the number of the line where you see the pencil.
4. Now, close your right eye and look at the pencil with your left eye, being careful not to move your head. Make another mark for where you see the pencil with your left eye. Repeat steps 3 and 4 for each distance.

   The change in position of the object as seen against the lines will be called "parallax." For example, if you see the pencil in front of line number 3 with your right eye, but it is in front of line number 9 when looking with your left eye, this is a parallax of $9 - 3 = 6$.

   A partially completed data table might look like the one below.

DATA TABLE EXAMPLE

Distance from pencil	1	2	3	4	5	6	7	8	9	10	11	12	13	14	15	16	17	18	19	20	Parallax
1 meter			I						I												6 spaces

Experiment 2—In this part of the activity, you will determine the effect on parallax of the distance between the two points of observation—the baseline.

FIGURE 2

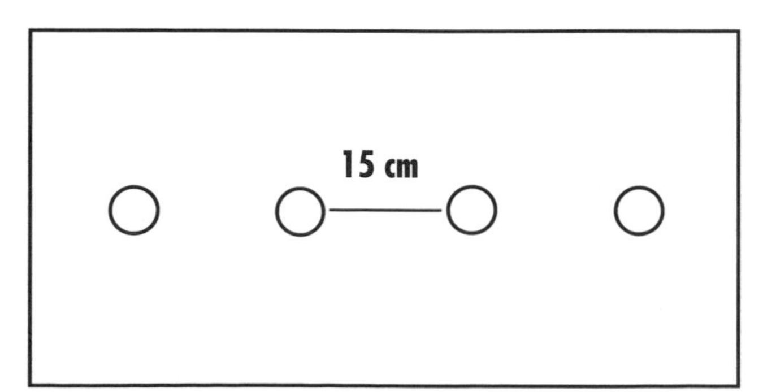

1. Find a piece of cardboard or stiff construction paper and punch four holes in it 15 cm apart as shown in figure 2. A manila folder torn in half at the fold and taped end to end works well for this.
2. Have someone hold this paper at eye level for you in one place 3 m from the chalkboard. This paper should not move.

You will be making all of your observations from the same distance from the chalkboard.

3. Again, have someone hold the pencil for you in front of the chalkboard with the lines on it, but this time have them hold it 1.8 m from the board.
4. Look at the pencil against the lines through the first hole with one eye and then through the second hole with the same eye. Record the parallax in Data Table 2.
5. Repeat step 4 for the first and third holes and for the first and fourth holes. Record the parallax each time in Data Table 2.

DATA TABLE 1

Distance from pencil	1	2	3	4	5	6	7	8	9	10	11	12	13	14	15	16	17	18	19	20	Parallax
1 meter																					
2 meters																					
3 meters																					
4 meters																					

DATA TABLE 2

Baseline	1	2	3	4	5	6	7	8	9	10	11	12	13	14	15	16	17	18	19	20	Parallax
15 cm																					
30 cm																					
45 cm																					

Questions and Conclusions

1. What is the effect on parallax of increased distance to the object being observed?
2. Is it possible that you could move so far away that there would be no parallax?

3. What is the effect of an increased baseline on parallax?
4. What would happen to parallax if you:
   a. increased baseline and decreased distance to the object?
   b. decreased baseline and increased distance to the object?
   c. increased baseline and increased distance to the object?
   d. decreased baseline and decreased distance to the object?
5. Hold a sharpened pencil in each hand at arm's length and point them toward each other. With both eyes open, try to touch the points of the pencils together. Now close one eye and try to touch the points again. What happened? Why?

How Far to the Star?

What Is Happening?

Measurements of distance are made either directly or indirectly. When a meter stick or similar instrument is held to an object in order to measure it, a direct measurement is being made. However, when objects are too far away to reach or too big to handle, indirect measurements can be made. One type of indirect measurement of distance takes advantage of the **parallax effect**.

The parallax effect is a phenomenon in which an object being observed against a stationary background appears to move when observed from two different points. The effect is easily demonstrated by lining up your thumb with some object in the distance, viewing your thumb with one eye and then with the other. The thumb appears to move against the background. The brain uses the parallax effect in depth perception. This is possible because humans have two eyes set apart by some distance.

The two main factors that determine the magnitude of the parallax effect are the distance to the object being observed and the distance between the two points of observation, known as the baseline. As distance to the object increases, parallax decreases. For this reason, parallax is helpful in observing the closest stars. As baseline increases, parallax increases. The largest baseline achievable on Earth is the diameter of Earth's orbit around the Sun. To do this, one observation is made, and then a second observation is made six months later. This largest baseline is necessary to measure the distance to the closest stars.

In this activity, students will investigate the effect of distance to the object and baseline on parallax.

Important Points for Students to Understand

◊ Distances and sizes in astronomy are often too great to measure directly.

◊ Parallax is one form of indirect measurement.

◊ Parallax is the apparent shifting of an object when viewed against a stationary background from two different points.

◊ Parallax can be used to measure the distance to objects that are too far to measure directly.

◊ The magnitude of the parallax effect depends on two factors—distance to the object being observed and distance between the two points of observation (baseline).

Materials

For each group of students:

◊ construction paper or manila folder

◊ pencil

◊ metric ruler

◊ chalkboard or, as an alternative, large sheets of paper (69 cm X 81 cm)

◊ chalk

◊ scissors (if manila folders are used)

◊ tape (if manila folders are used)

◊ single hole punch

Vocabulary

Direct measurement: A measurement made using an instrument applied directly to the object or distance being measured.

Indirect measurement: A measurement made by other than direct means. Physical proximity to the object or distance is not necessary. Examples of indirect measurement include angular diameter and parallax.

Angular diameter: The measure of an object's diameter in degrees of an arc rather than in linear measurement units. This is a type of indirect measurement. Angular diameters are used to calculate diameters of objects in the Solar System.

Parallax effect: Usually referred to as "parallax," it is a phenomenon in which an object appears to change positions against a background when viewed from two different points.

Baseline: The distance between the two points of observation when measuring parallax.

Time Management

This activity will take one class period or less. It could be spread out over two days in a couple of ways. First, if there is only one chalkboard in the room, it will be difficult for all students to participate at once. Alternatively, the students could make their twenty lines on large sheets of paper and fasten these to the wall. This could be done on the day prior to the activity. Second, the pieces of stiff paper with four holes in them could be constructed ahead of time. If both of these things are done, taking the measurements should consume less than half a class period.

Preparation

Be sure that the room is large enough for all students to work at the same time. It is helpful to mark in advance the spaces where students or groups of students can work. An alternative to having students draw the lines on the board is to draw them on adding machine tape and then attach it to the wall. This can be done prior to the activity. Have all materials ready and centrally located for distribution. Students will need to work in groups of three in this activity. Group them before class to save time.

For further information, refer to Reading 3, "Hubble Space Telescope;" and Reading 6, "The Parallax Effect."

Suggestions for Further Study

Students may want to determine the actual distance at which they can no longer detect a parallax with just the distance between their eyes as the baseline. They may also like to experiment outdoors with larger distances and baselines. Rather than using lines on the chalkboard, they can use trees or other landmarks as their background.

Parallax is not used nearly as much today in determining distances to objects in the Solar System as it was years ago, but is still sometimes used to find distances to stars, both from ground-based telescopes and from space missions. Light has replaced the parallax method for measuring these distances. Students might find it interesting to research how this is done.

There is another way to demonstrate parallax that involves the students in a whole-class activity. It is also a model of the way parallax is used to measure the distance to stars. Take the whole class outside in a large open space—a soccer field, for example. Arrange them as shown in figure 3. One student is designated as the near star which will be observed. About half the class stands three or four meters behind this student representing the distant stars. They will be the stationary background against which the near star will be observed. Fifteen to twenty meters on the other side of the "near star," the teacher stands with the other half of the class arranged in a circle around him or her. The teacher represents the Sun and the students represent Earth at different positions in its orbit around the Sun. Have the students walk

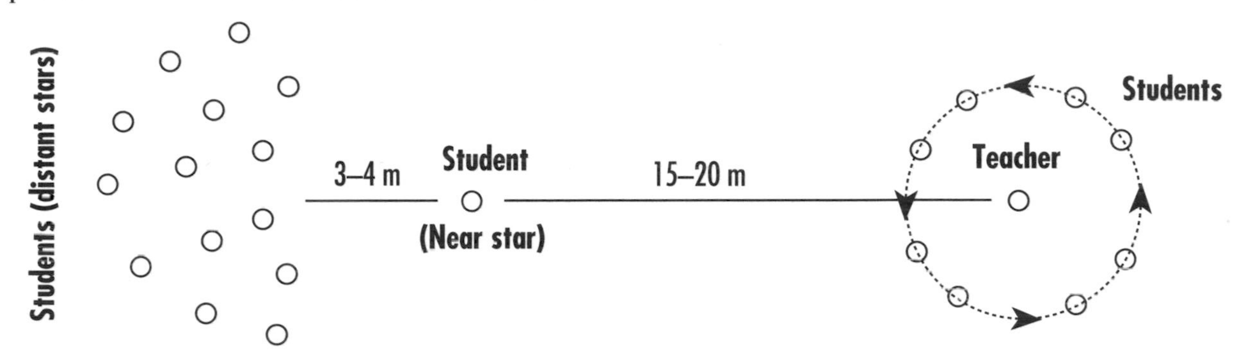

FIGURE 3

around the teacher in a circle and make observations of the near star. Ask them to concentrate on how the near star appears against the background stars from different points in the circle.

Answers to Questions for Students

1. As distance to the object increases, parallax decreases.
2. The obvious and acceptable answer is "yes." The correct answer is actually "no." The parallax is always there. It becomes too small for us to detect it, but it is always there.
3. As baseline increases, parallax increases.
4. a. Parallax would increase.

   b. Parallax would decrease.

   c. The answer depends on how much each factor was changed. The way the two factors are being changed, they offset each other. Therefore, parallax may increase, decrease, or stay the same.

   d. The same answer as c.
5. The students should not be able to touch the pencil tips together with one eye closed. The reason: with one eye closed, there is only one point of observation and the brain cannot take advantage of parallax. Therefore, it is much more difficult to tell which pencil is closer and which is farther away. This makes it extremely difficult to touch their points.

Stars

Alone in the night
  On a dark hill
With pines around me
  Spicy and still,

And a heaven full of stars
  Over my head,
White and topaz
  And misty red;

Myriads with beating
  Hearts of fire
That aeons
  Cannot vex or tire;

Up the dome of heaven
  Like a great hill,
I watch them marching
  Stately and still,

And I know that I
  Am honored to be
Witness
  Of so much majesty.

Sara Teasdale

Solar System Soup: The Formation of the Solar System

Background

Over the last four centuries, many theories have been formulated to explain the origin and evolution of the Solar System. Today, the theory most commonly held by scientists is known as the **accretion theory**. The theory was originally formulated in the 1940s but it has been refined over the last few decades.

The theory explains that the Solar System began to form at a time when the Sun had acquired enough mass to begin to attract material from what is known as the **interstellar medium**. The interstellar medium is simply the gas and dust that are spread throughout our galaxy. Since the Sun was spinning when it began to attract the gas and dust, what originally formed was a huge rotating disc of material with the Sun at its center. Astronomers believe that this disc looked very much like the satellite pictures of hurricanes which you have probably seen on television weather reports. Over billions of years, the material in the disc accreted, or clumped together, to form the planets presently in the Solar System. Prior to the accretion theory, no one could explain why the planets farther away from the Sun revolved more slowly than the ones closer to the center of the Solar System.

This activity will model what the very early Solar System would have looked like, as predicted by the accretion theory, and demonstrate that the material on the edge of the disc revolves more slowly than the material at the center.

Objective

The object of this activity is to observe a model of how the Solar System would have originated according to the accretion theory.

Materials

For each group of students:

- ◊ 1 stirring rod
- ◊ 1 bucket (11 L) for student exercise
- ◊ basin for collecting used water
- ◊ 15 mL of vermiculite

Procedure

1. Fill a bucket (or comparable container) three-fourths full with water.
2. Using the graduated cylinder, measure about 15 mL of vermiculite and pour it on top of the water.
3. Stir the mixture vigorously with a stirring rod in a circular motion. When you have a funnel-shaped pattern in the water, stop, remove the stirring rod, and observe.
4. In the space provided on the following page, sketch the

pattern of vermiculite that you observe. Label what parts of the pattern might form the inner planets and outer planets of a Solar System.

Sketch of pattern of vermiculite in bucket

Questions and Conclusions

1. What did you observe happening to the vermiculite once you stopped stirring the water?
2. Describe the pattern of vermiculite on the surface of the water.
3. Did all of the vermiculite spin around in the bucket at the same speed? If not, what parts spun faster and what parts spun more slowly?
4. Why did the vermiculite eventually slow down and stop? Why has the same thing not happened to the Solar System?
5. Think about how big your bucket of water and vermiculite is compared to the size of the Solar System. Why do you think it took millions of years for the Solar System to form? Can you think of reasons other than size?

Solar System Soup: The Formation of the Solar System

What Is Happening?

Although many theories have been formulated to explain the origin and formation of the Solar System, the one most commonly held among astronomers is the **accretion theory**. This theory maintains that the Solar System began to form when a rotating interstellar cloud of gas and dust collapsed under its own gravity. This **protosun** was already rotating. Therefore, as it collapsed, it began to rotate faster and consequently formed a huge flat disc of gas and dust. The gas and dust were not evenly distributed, but were in disconnected parts throughout the disc. The German physicist Baron Carl Friedrich von Weiszacker proved that in such a disc, the outer parts would revolve around the center of the disc more slowly than the inner parts. Later work showed that, over time, the disc would become even more clumpy (disconnected), and eventually these clumps would become dense enough to fall together under their own gravitation. The clumps then evolved into the planets, according to the accretion theory.

One problem with the accretion theory is that the Sun has been found to rotate more slowly than the theory would predict. If the disc spins faster as it collapses, the theory predicts that the Sun should spin faster than is actually observed. This problem was solved in the early 1960s with the discovery of the **solar wind**. This is the stream of charged particles that the Sun is constantly emitting. The Sun's magnetic field is like the spokes of a bicycle wheel carrying streams of charged particles along with it as it spins. This slows the Sun down. This phenomenon, referred to as **magnetic braking**, is the reason why the planets now have most of the rotational momentum in the Solar System rather than the Sun itself.

This activity is designed to show students the pattern that is formed when disconnected material begins to spin. They will see how the outer parts of the spinning disc revolve relatively slowly, and they may even see some of the material begin to clump together into "planets." Be sure the students understand that, unlike the theorized "clumping" of planets due to gravity, the

Materials

For the class:

- ◊ 1 bag of vermiculite
- ◊ containers of water placed centrally in the room
- ◊ 100-mL graduated cylinder
- ◊ 1000-mL beaker or comparable container for overhead demonstration, if desired

For each group of students:

- ◊ 1 stirring rod
- ◊ 1 bucket (11 liters) for student exercise
- ◊ basin for collecting used water
- ◊ 15 mL of vermiculite

Vocabulary

Accretion theory: A theory formulated in the 1940s by astronomers to explain the formation of our Solar System. According to this theory, material in a disc rotating around a newly formed sun accreted or clumped together to form the planets.

Interstellar medium: The gas and dust spread throughout interstellar space. It is concentrated in a spiral arm pattern.

Protosun: An early stage of star formation in which a large cloud of interstellar medium is contracting under its own gravity.

Solar wind: A stream of electrically charged particles that is constantly flowing outward from the Sun. Near Earth it is traveling at about 600 km/s.

Magnetic braking: The slowing of a star or planet's rotation due to the interaction between its magnetic field and surrounding ionized material.

clumps they observe in the vermiculite are *not* due to gravity, but to cohesion of the vermiculite.

Important Points for Students to Understand

◊ While many theories concerning the origin of the Solar System exist, the one most commonly held by scientists is the accretion theory.

◊ The disc from which the Solar System formed was not solid like a discus. It was made of disconnected material as demonstrated in this activity.

◊ In a disc such as the one from which the planets formed, the outer parts of the disc revolve more slowly than the inner parts.

◊ Material rotating in a disc such as the one the Solar System formed from will eventually form clumps due to gravity and the differential speed of revolution. The clumps then become massive enough to attract other material gravitationally, forming a sphere.

Time Management

This activity will take less than one class period since it does not require data collection and does not take long to set up. In order for the activity to take even less time, it may be done as a teacher demonstration by simply placing a beaker containing water and vermiculite on an overhead projector. The image of the rotating disc will be projected on the screen, and you can point out the relevant features.

Preparation

Prior to the class, place the buckets, graduated cylinder, stirring rods, and vermiculite in a central location. Vermiculite can be obtained from most garden stores. Have the students come and collect their materials, measuring out the vermiculite before returning to their seats. Ideally, there would be a sink in the room from which to get water. Alternatively, the teacher may fill the buckets before class or have a large enough container of water to supply all the groups.

For additional information, refer to Reading 5, "Scouting Earth /Moon."

Suggestions for Further Study

While the accretion theory of the origin of the Solar System is the predominant one today, this has not always been the case. Other recorded theories date back to the 1600s. Students may be interested in investigating some of these. Many introductory astronomy books contain some detail about early and competing theories. Studying these theories could lead easily into a discussion of Kepler's laws.

One drawback of this experiment is that the disc eventually stops rotating unless stirred periodically. Such stirring, however, disturbs the phenomenon being observed. A solution is to place a magnetic stir bar on the bottom of the beaker or bucket and use the type of hot plate that also stirs. Using heat is not required. Such an arrangement allows students to observe what happens to the vermiculite over an extended period of time.

Answers to Questions for Students

1. Students should see a pattern form in the vermiculite that resembles a satellite photograph of a hurricane. They will also see the disc of material slow down and stop.
2. The vermiculite will probably not be evenly distributed in the disc. Most of it will be concentrated in the center, becoming less and less dense toward the outer edge of the disc.
3. All the vermiculite should not spin at the same speed. Students should see the inner parts of the disc spinning more rapidly than the outer parts.
4. The vermiculite is floating on water. The water is in contact with the sides and bottom of the bucket creating friction. This friction causes the water and, therefore, the disc to stop spinning. Wind resistance is also a factor but not nearly as important as the friction between water and bucket. The same thing does not happen in the Solar System because there is essentially no friction between the planets and anything else.
5. Finding the answer to this question will require some thought and research on the part of the students. Size, thin distribution of matter, and opposing gravitational forces are some factors that could have influenced the formation of the Solar System.

Red Sky

Red sky at night, sailors delight
Red sky in the morning, sailor take warning.

Evening red and morning gray
Sets the traveler on his way;
Evening gray and morning red
Brings down rain upon his head.

(folklore)

The Goldilocks Effect or "This Planet Is Just Right!"

Background

Earth is the only planet in the Solar System that supports life. Neither Venus nor Mars, Earth's closest neighbors, is habitable. One of them is far too hot and one is far too cold. Mars has practically no oxygen in the atmosphere and most of the water is tied up in polar ice caps and permafrost. On Venus, the temperature stays around 454° Celsius. The highest temperature ever recorded on Earth was 58° Celsius in El Aziza, Libya on September 13, 1922. If Earth were either closer or farther away, life as we know it never would have evolved.

In this activity, you will investigate the way distance from a light source affects temperature—one of the many reasons why Earth is "just right" in its ability to support life.

Objective

The objective of this activity is to investigate the relationship between distance from a light source and temperature and to apply this relationship in understanding why life is possible only on Earth in our Solar System.

Materials

For each group of students:

- 4 Celsius thermometers
- meter stick
- reflector lamp (equipped with clamp) or gooseneck lamp with 75-watt bulb
- watch or clock

Procedure

1. In this activity you will find where on the meter stick a thermometer should be placed so that the temperature it measures matches a target temperature your teacher specifies before the experiment begins.
2. Place the meter stick on a table. Place the lamp at the end of the meter stick so that light is aimed down along the meter stick as shown in figure 1.

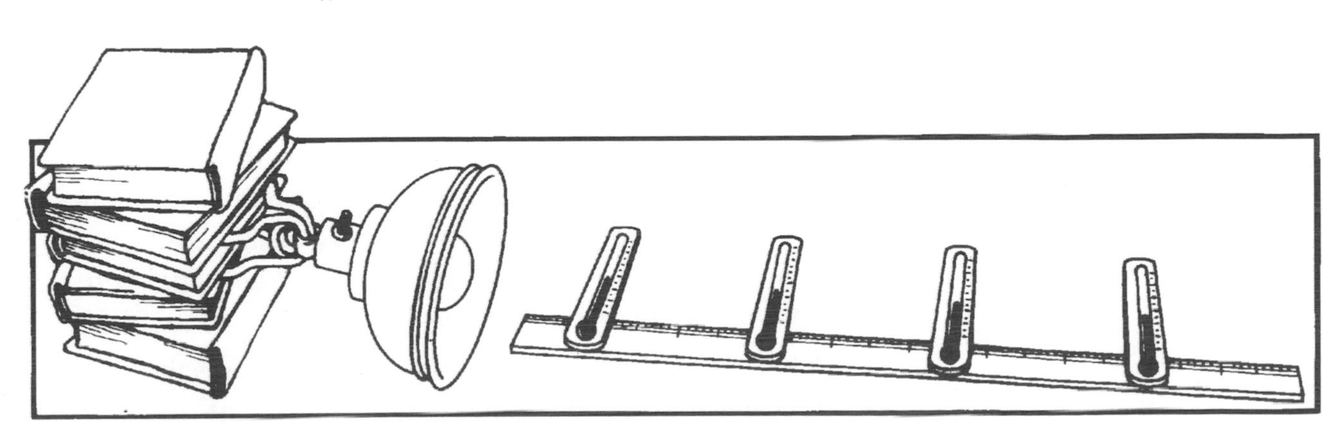

FIGURE 1

3. Place the thermometers down flat on the meter stick in positions where you think one of them will have a tempera-

ture that will match the target temperature. Each thermometer represents a possible distance for Earth from the Sun. Be sure that the bulb of each thermometer is on the meter stick. Record the distance of each thermometer from the end of the meter stick that is next to the lamp in the Data Table.

DATA TABLE

Trial	Thermometer	Distance to thermometer	Temperature (° C)					
			Start	3 min.	6 min.	9 min.	12 min.	15 min.
1	1							
	2							
	3							
	4							
2	1							
	2							
	3							
	4							
3	1							
	2							
	3							
	4							

4. Record the starting temperatures for each thermometer in the Data Table. Turn on the lamp and record temperatures for each thermometer every 3 minutes until no temperature change is seen in any thermometer. ***Caution: The lamp and reflector will become hot.***
5. If none of the final temperatures are the same as the target temperature, graph your results, plotting temperature versus distance. Then try to predict at what distance you should place thermometers when you repeat step 4 so that one of the final temperatures matches the target temperature. If